EXUBERANCE

Also by Dolores Hayden

POETRY

American Yard

Nymph, Dun, and Spinner

NON-FICTION

Seven American Utopias

The Grand Domestic Revolution

Redesigning the American Dream

The Power of Place

Building Suburbia

A Field Guide to Sprawl

EXUBERANCE

poems

Dolores Hayden

🐓 Red Hen Press | *Pasadena, CA*

Book design by Mark E. Cull
Cover Photograph: Flight demonstration by Orville Wright, Berlin, 1909. Sueddeutsche Zeilung Photo/Alamy Stock Photo

Library of Congress Cataloging-in-Publication Data
Names: Hayden, Dolores, author.
Title: Exuberance / Dolores Hayden.
Description: First edition. | Pasadena, CA : Red Hen Press, [2019]
Identifiers: LCCN 2018055876 | ISBN 9781597096041 (tradebook)
Classification: LCC PS3608.A916 A6 2019 | DDC 811/.6—dc23
LC record available at https://lccn.loc.gov/2018055876

The National Endowment for the Arts, the Los Angeles County Arts Commission, the Ahmanson Foundation, the Dwight Stuart Youth Fund, the Max Factor Family Foundation, the Pasadena Tournament of Roses Foundation, the Pasadena Arts & Culture Commission and the City of Pasadena Cultural Affairs Division, the City of Los Angeles Department of Cultural Affairs, the Audrey & Sydney Irmas Charitable Foundation, the Kinder Morgan Foundation, the Allergan Foundation, and the Riordan Foundation partially support Red Hen Press.

First Edition
Published by Red Hen Press
www.redhen.org

The objects on the ground now seem to be moving
at much higher speed, though
you perceive no change in the pressure
of the wind on your face. You know then that
you are traveling with the wind.
                    —Orville and Wilbur Wright, 1908

Who can do anything better than this propeller?
                    —Marcel Duchamp, 1912

Contents

III

EXUBERANCE

Kitty Hawk, 1900

Addie Tate

Wilbur unrolled white French sateen
cut on the bias for strength, lined

with fifteen narrow pockets for the ash ribs.
He borrowed my sewing machine,

shortened his cloth to match the white pine spars.
We came down here for wind and sand,

and we have got them, said Orville.
Everything from dead calm to a whole gale.

In twelve attempts one day, Wilbur spent
barely two minutes in the air. Late October,

the weather turned. Discarding the glider,
he promised to return next summer

with a new machine. So, he told us,
salvage anything you want. I stayed up late

running the treadle, stitching Sunday dresses
for Irene and Pauline. They shimmered

when they wore those wings. The ribs and spars
their father fed to the fire that winter.

I

Dominguez Field Air Meet

Champ Pickens

I'm the devil's own publicist, promoter,
twenty-four sheet billposter.
Hell's mint! I am prince of pitch,
god of gab for a week of wings.

Call me champion of ballyhoo,
I invent the event. Boldface you.

Once I raced bicycles hell-to-toot,
no money there—but airplanes!
What a cocktail, half a dozen flying machines
fizzing above full grandstands:
seats, fifty cents; box seats, one dollar;
standing room, a quarter; children, ten cents.

Lincoln Beachey, the Boy Aeronaut,
is steering a dirigible,
he's never seen a plane.

Finger the canvas, I tell him, stroke
the struts, pry apart the engine, oil it.

A flying machine throbs, kicks.

Mount her, gallop into the game.
Grab the stick, wink at farmers
who wish they could fly,
take off with an ear-pounding roar.
Thousands will pay to watch you
wrestle those wings,
yell as you roll, pitch, yaw.

I front so many pilots,
they call me the Champ,
call me the banker
who bags the long green.

My cut is a hell-whooping half.

I tell the time with solid gold, I strut
in a suit and spats. Watch me ballyhoo
from hell to breakfast, watch me spiel
and grind till the last ticket's torn.

Bird Man

Linc Beachey, pilot

Left school at twelve, can't follow physics,
I'm just a man stepping out on the law
of gravity, making my living leaving—

and reappearing. Morning, evening,
I am The Man Who Owns the Sky.
I fly over Niagara, down into the mist,

close to the whirlpools—whoooooeeee!—dart
under Honeymoon Bridge, soak my suit
as wheels kiss water. You call it suicide?

Risk improves my mood and the money's good.
I burn my last drop of fuel to climb two miles
above Chicago, ride a dead stick all the way

down, proclaim the gospel of wings—
heaven pricked by human invention.
I spin a holy ruckus in the sky,

racket my blessing above root beer, red hots, popcorn.
Give me a hundred thousand pairs of uplifted eyes.
Amen.

Bird Woman

Betty Scott, pilot

The Wright brothers warped wings
to corner the wind. They built a seat
for racing, wrapped
their flight school with rules:

No drinking.
No swearing.
No flying on Sundays.
No females.

What man claims God does not allow women to fly?

Glenn Curtiss says he won't teach women,
either. But I know motors: my father gave me
the keys to the Cadillac when I was thirteen.

To keep me out of trouble, he said.

Last year I coaxed an Overland coast to coast.
On the door: The Car, The Girl, and The Wide, Wide World.

Yes, I'm a finishing-school girl,
Miss Blanche Stuart Scott of Rochester,
five foot one, dark red hair.
If I marry myself
to a rackety engine,
of course, it will attract publicity.

I partner my plane in a ballroom of air.
Up where the landscape opens, I steer
without rutted roads to slow me.

Clouds nudge my feet as they shadow-slip
across lakes and farms two hundred yards below.

Flying Lesson: Clouds

Focus on the shapes: *cirrus*, a curl,
stratus, a layer, *cumulus*, a heap.

Humilis, a small cloud,
cumulus humilis, a fine day to fly.

Incus, the anvil, stay grounded.
Nimbus, rain, be careful,

don't take off near *nimbostratus*,
a shapeless layer

of rain, hail, ice, or snow.
Ice weighs on the blades

of your propeller, weighs
on the entering edge

of your wings. Read a cloud,
decode it, a dense, chilly mass

can shift, flood with light.
Watch for clouds closing under you:

the sky opens in a breath,
shuts in a heartbeat.

Suiting Up

Harriet Quimby, pilot

Leslie's Weekly sent me "Around the World with a Camera."
Back in New York, I wanted to cover aviation.
My editor shrugged. Your neck, Harriet.

I enrolled in a school of sky to scoop New York,
learned to balance silk, bamboo, and steel,
shift my weight,

shape the arc, soar and swoop.
My costume? Leather gauntlets, a violet satin flight suit
with trousers I tuck into laced-up leather boots.

Father swears I no longer look like a respectable woman.
He sells Quimby's Liver Invigorator door-to-door,
grumbles it's unnatural to fly.

Mother winks. She knows I can button
my wide purple trousers
back into a long skirt if I choose propriety.

With a pilot's license in my hip pocket,
I cross the Channel
in my Blériot: zero-zero, no visibility

in the fog, goggles so wet I can barely see.
I have the air intoxication,
and only a flier knows what that means.

Risk

Betty

Pre-flight checks? I do them and
I do them again, every wire and line,
every bolt and screw. Done. And done.

I leave the fear behind like a pair
of tight kid gloves I toss as I taxi down the grass,
what if, what if, what if—wheels lift,

fields drop, roofs flatten,
and I'm weightless, bare-handed,
engine noise deadens thought.

To twist into the sky, I gun the motor,
nose up, nose up,
then over, over and down, gravity

presses my chest, I dive
and roll, roll again, tease
the crowd, wield wings this way, that way,

I'm low enough for them
to spot my grin
before I land inside a painted circle

in front of the grandstand. Perfect.
Out of the cockpit, dizzy, oxygen-
deprived, half-deaf, I hear ten thousand shout

as if summer thunder rumbled. I'm all adrenaline.
It is not about the trophy or the cash.
My body begs my brain for speed and stunts and sky.

Drag

Linc

Why does any woman want to fly?
Betty and Harriet pull as much press
as those pushy suffragettes.

At an LA air meet, I appear as "Florence."
In a cloche and a dress, I splash around in the sky,
one near miss after another—the worst
female pilot anyone has ever seen—
and the most exciting stunt.

I turn up in Chicago as "Clarice," veer
toward the grandstands as spectators shriek, bounce
down a city street as people scurry, zoom
above Lake Michigan, drop on a ferry
as passengers dive off the deck.

Glenn Curtiss grins: That cub, Beachey. . . .
Champ Pickens shakes his head: Helluva show.

Landscapes

Harriet

I lift off above leaf-green spring, taut
tennis courts, lazy fairways, lines
of narrow sailboats moored
where the sea shifts blue-green, gray-blue.

I level out above circles of pear blossom,
grids of corn, pastures patterned
with weathervane cows, heads to the wind.
A breeze passes across every grassy hill,

ripples every light-struck river. It frays
and feathers cirrus geographies
I skim, skip through, explore
to pierce a pantheon of old superstitions,

drive my shadow across the land,
reorder the mechanics of awe.
Sitting in the sky, I scout high prairies
of air, yearn to settle there.

Daredevil

Champ

An air devil perfects a trick, confidence
multiplies confidence. A rival wills a wilder
stunt, a hell-all gamble just to stay in the game.

I get a cut of everything
and men will bet on anything—

mechanics prime motors with castor oil,
spectators scent sudden death,
bookmakers lay odds on the day a pilot

dies in the Undertaker's Chair.
Three times my lollapalooza earner quits.

The Man Who Owns the Sky plans to play
the stock market, woo some women.
I tell him an air devil never retires unless

someone invented a stunt he can't do.
Linc mounts the cockpit in a suit and tie,

reverses his cap to signal the start:
flip, loop, slide, drop, he delivers dare
after dare after dare after dare.

Love Life

Linc

After I fly the Ocean Roll
and the Corkscrew Twist,
women come to my hotel.

I'm never dizzy,
I don't drink,
I don't smoke,
I carry diamond rings.

Some women want a proposal
before we Loop the Loop.

Flying Lesson: Winds

Always take off into the wind.
Call the wind Conductor, Doctor,

compliment the way it smoothes and soothes.
Warm fronts shift, cold fronts shift

faster, watch the leading edge. Measure
wind speed with your eyes:

smoke drifts in light air; leaves
rustle in a light breeze; flags

flap in a gentle breeze; flags
snap in a moderate breeze; saplings

sway in a fresh breeze.
Hangar your plane in a fresh gale.

Curse the wind: Barber, Bricklayer.
Clock how it whistles, whines

through your struts.
Learn all the names:

Blue Norther, Santa Ana.
Memorize the maps: winds scribble

their signatures over and over.
Always land into the wind.

Accident

Betty

Harriet circles a lighthouse outside Boston.
A sudden gust tips the tail, and she drops

from the cockpit in her purple suit,
tumbles head first as her Blériot
glides into the marsh.

Sun lacquers its wings.

Reporters ask: As a woman
who has just watched
another woman die, will I stop flying?

Certainly not.

Linc says Harriet must have fainted,
going 100 miles an hour,
speeding down from 5,000 feet,
a light, delicate woman,
the terrific rush of air
was too much for her.

Fainted? Take that back.

Collar and Tie

Champ

Some ticket holders salivate for a scorched fuselage.
They sprint across the field, seize
fragments of an engine, twisted wires,
torn cloth, all the splinters of aspiration.

Men—and women, too—will grab
a bloody cap, leather gloves, a collar and tie,
carry them off as if the flier were a god
and these were disguises, everyday,
earthly things worn in the sky.

If a pilot dies, I remove the body,
burn wreckage on the spot.

Hell's delight, the gate doubles the next day.

Postures

Linc

The Wright brothers weave nests
of wood and cloth and wire.
They place the bird next to the engine.

Glenn Curtiss seats the bird
in front of the engine.

Wilbur, Orville, and Glenn can't agree
on anything except—

I am the greatest bird anyone has ever seen.

I wear the plane.

Tell me roll,
tell me dive,
tell me breathe,
tell me high,

I steer with hands or knees,
I'm first, fastest, highest, farthest,
and I'm ordering a machine
with an aluminum skin, Beachey,
B-E-A-C-H-E-Y across the wings.

Glenn doubts the strength of the alloy.
He's been saying things like that for years.

Flying Lesson: The Aviation

Mix two ounces gin,
one ounce lemon juice,

two dashes maraschino,
two dashes crème de violette:

sky-blue cocktail, cold
as an open cockpit.

Packing the Parachute

Tiny Broadwick, parachute jumper

Air circuses always put the jump last—
sometimes a harness breaks
or a gust shoves you sideways
like a liquored-up husband.

Wilbur Wright shook my hand once:
You're awful small to do that.

Honey, I'm eighty pounds, fifty-some inches,
I've landed in the chill of Lake Michigan
and the hot steam of a moving train.
Anything is easier than thirteen-hour days
in a North Carolina cotton mill.

I married at twelve and had a baby. Honey,
that's the way it was done down South.

I left Verla with my mother. At fourteen, I joined
the Johnny J. Jones Carnival in Jacksonville
and did all I ever wanted to do:
jump from a hot air balloon filled with gas.

I was The Doll Girl in white ruffles and a bonnet.
I dropped like a baby falling from a bassinet.

On the Fourth of July I stunted
with red, white, and blue chutes.
Faster than a firecracker,
I cut each set of cords, fell free
before opening the next,
and landed on a trapeze by the Stars and Stripes.

An airplane is much safer than a balloon.

I pack every inch of my parachute
flat and dry. Fold, unfold, fold again,
life or death in a thousand jumps.

I drop from a plane cruising at seventy.

Silk like a coverlet,
rope like an umbilical cord,
a parachute clasps birth, rebirth, in a bundle.

Watch gravity work,
watch me fall,
watch me
watch

before the chute snaps open.

Panama–Pacific

Champ

San Francisco fairgoers jam grandstands, lean
from open windows, crowd onto roofs, tip
their faces up: Lincoln Beachey

taxis out from the Palace of Machinery,
takes off to applause, lands to ovations.

He performs the Twentieth Century for them.

Officials gather to bestow a medal—
but the medal is not here yet.

Could he go up one more time?

He scrawls autographs, poses for photographs,
jump ropes twirl and children chant:

Lincoln Beachey
thought it was a dream,
soaring up to heaven
in a flying machine . . .

Could he go up one more time?

He climbs in the cockpit, heads out
above the bay, spins through loop
after loop, circles up, loops again, dives.

Halfway down his left wing shears off.
And then—his right.

Hell and damnation,
he's gone.

II

The Wing Walker

Champ

Lieutenant Ormer Locklear scrambles
out of his cockpit, his partner grabs
the stick as Lock swings up onto the wing.

Holy moley, he's out there dancing
like a bear in slick-soled boots—
he's hanging by his knees
from the undercarriage, he's astride
the tail, waving to the crowd,

he's changing from plane to plane
in mid-air. Hell-in-harness, what an act!

Bill this as a circus, a flying circus
three miles long and a mile high,
Lock with his pals, Skeets and Shorty.

I tell them, turn up in uniform. Women
will go crazy for daredevils
with wings and ribbons on their suits.

Stunt Money

Lock Locklear, pilot

Records are set and broken, reset,
broken again, just like fliers' bones.

Men die learning to fly here in Texas.
And the lifespan of an army pilot overseas?
Two weeks. Maybe three.

To cheer up recruits, I stroll the upper wing,
do a couple of handstands.

It's just a gag until Champ wants me to wing walk,
Uniontown, Erie, Atlantic City,
rain, shine, or cyclone, a thousand dollars a show.

I sign. When I slip, Champ grins:
Bandages are box office.

The Barnstormer

Pang Pangborn, pilot

Left the army, bought a surplus Jenny,
you bet the ship is a bit of a crate,

no wheel brakes, bungee cords
for shock absorbers. The engine
emits a muted whistling clatter
from the valve springs, overheats,
spews fumes and oil in my face,

but I'm earning a living playing the pastures.

I hedge-hop into a town, buzz some barns,
land in a flat field. Rides, $3.00.
Long Rides, $5.00.

Locals with no cash trade eggs
or a couple of chickens.

Rides are routine. Flying upside down,
if the fuel fails to flow, I stall—
and a stall could be fatal—
split second by split
second, I baptize my act.

At night I curl into a hammock
slung under a wing, count the take.

Next morning, I follow a road or a railroad line
to another town, greet another set of folks
who have never seen an airplane.

Let's get acquainted!

Flying Lesson: Strange Field Landings

Strap on the map, a strip of landscape
scrolls on your knee board:

climb north from town through a gap
in the ridge, cross the switchback

on the mountain. Pick up
the west branch of the river—

look sharp, five small houses,
follow the white gravel road.

South of the next town stretch pastures
you can land on in a pinch if you need

to get down, but take good care,
from a thousand feet up, brooks disappear,

gullies vanish. Knolls flatten; marshes
green up like meadows. Watch out

for wires, telephone poles, tall trees.
Beware a herd of sleepy cows,

keep an eye skinned
for a flock of muddy sheep.

Feel

Ruth Law, pilot

Yesterday a rainbow wrapped my cockpit.

When I'm not flying, I'm planning to fly,
scribbling compass headings
on my cuff, counting records I set today,
records I might seize tomorrow.

To balance eight hundred pounds of plane
by feel and fuel gauge, I travel past jitters.
My ship wavers, shivers,
the scare is part of the thrill.

I sleep out on my hotel roof in November,
harden myself to a high, cold sky.

Then I brace for tailwinds, grip the controls.

With the wind behind me, I head east nonstop
at over a hundred miles an hour,
Chicago, Illinois, to Hornell, New York,
and I break the distance record—not
the women's record, the American record—

five hundred and ninety miles. Done.

Shroud Lines

Tiny

Strong sun. Crowds stir
as engines mount
and fade, snarl and whine.

I wear a short white dress.
I have packed my parachute perfectly.

At the end of the afternoon, Pangborn carries me up,
and I move out onto the wing,
skirt fluttering.
I calculate wind speed,
drop,

and the parachute starts to fill.
I drift down quietly,
slipping the shroud lines

to land precisely on the grass.

The crowd surges forward. The silk canopy
rustles, whispers to those close by me.
I have entered the empire of the air
and returned to them, intact.

Heaven

Pang

Speeding to Alabama, I lose
my bearings, land in a pasture
somewhere in Mississippi.

Farmers from miles around arrive
in wooden oxcarts
to meet the man who can fly.

I give rides all day, lift them
into God's eye view—
cotton fields,
red dirt roads,
board-and-batten houses—

all of Oktibbeha County
opens before us,
the church steeple adorned
with a single cloud.

At dusk I cruise a slow sunset
as if I alone could make the moon appear.

An old woman whispers: Mr. Pangborn,
how much would you charge
to take me to heaven and leave me there?

Flying Lesson: Emergency Jump

Zero-zero:
no ceiling,
no visibility.

Five minutes' fuel.
Four.
Three.

Climb the cowling.
Bend your knees,

> hurl yourself
> head first
> into night and cloud.

>> Watch out for the ship
>> gliding down without you.

> Away from its path,
> jerk the cord.
> Release the silk.

> Cross your ankles.

Earth pulls you back,
hard and particular.

The Wing Walker in Hollywood

Lock

Mr. Ballyhoo banks half of every dollar I earn.
　　　Though he never climbs in the cockpit, Champ is my
personal capitalist. My promoter owns the plane, owns

　　one hundred sixty pounds of aerobatic me,

and after a dozen exhibitions, Champ sells me
　　　to Universal Studios—trick flyer, five ten, green eyes,
Clark Gable look-alike in a leather jacket.

　　More money for us, son, he says, but I can't act,

my film is stunts, crashes, kisses.
　　　Fans line up to watch me move, they wonder
how I stand tall in the wash from the propeller,

　　how I climb a rope ladder from a speeding car

to a moving plane, how I leap from one plane
　　　to another. (I never wear a parachute.)
For fun, I fly Buster Keaton right side up and upside down,

　　I scare Charlie Chaplin with a death dive,

I spin to tease Louise Lovely and Leatrice Joy.
　　　After I wing Viola Dana (adorable girl!)
down to the beach for a moonlight swim,

　　studio execs suspect I could be trouble.

Flying Lesson: Stunting

To stunt is to flirt: you beckon,
you spurn, you beckon.
You tune up a waltz, turn to a dirge,
swing back to a waltz.

You brag about how close you hold your partner.

Your engine hesitates—how many seconds?
Your machine misses an obstacle—how many inches?

It is fatal to be too casual, though casual
is what's demanded here,
no one wants to see the top attraction tremble.

Whatever the weather, people pay
to watch you fly,
not wait out the wind. They scream for you.
Know the limits of their patience. And yours.

Take the ship higher, faster.
Stress every inch of the ship, do things
the inventor never intended.

And then, do them upside down and backwards.

Lovebird

Champ

Hell's bells, Locklear's gone moony.
The Lovebird bounces his wheels
across Viola Dana's soundstage roof,
he wings her over Hollywood to drop
lipsticks, compacts, a slipstream of gilt,

a cataract of sky-swag for her fans.
Go bigger than lipsticks, I tell him.
Hell's wedding bells, get hitched,
marry that starlet in your plane, kiss her
in the air all over town, sell yourself

smooching. Snatch headlines. Fox
might raise your take. But first,
you have to divorce Ruby Graves,
your wife back in Fort Worth, the one
who tells you stunting is dangerous.

Safety Second

Lock

The Skywayman finishes with a night stunt,
a flaming tailspin, but Fox
demands a mock-up. Safer, they say.

       Safety second is my motto.

What are they paying me for? I invented
the wing walk, I switch from plane
to plane

       with a thousand feet of wind

singing between my body and the ground.
Pa called me his danger-loving boy.
Here I fly for the cameras:

       red filters turn day to night,

New York and Chicago are stages,
buildings fold
when stagehands pull a few ropes,

       shut the hinges, false front

after false front. If we're not shooting,
I loop, spin, wing walk over LA.
Dry grass, empty lots, unpaved streets

       in dead subdivisions

stretch not far from the homes of the stars.
In Texas I nailed houses,
watched speculators go under,

boom towns collapse into

ghost towns. Pilots inventory high disasters:
nosedives, tailspins, graveyard spirals,
doomsday spirals, crack-ups.

If prosperity turns out to be an illusion,

Wall Street will need new lingo
for an international bust-up.
Call it a bubble, a stumble, a dive?

Call it a Great Crash.

Flying Lesson: Fields and Flares

To illuminate an airfield at night,
the owner should hire a couple of boys

to line the runway with four-minute flares,
zinc buckets half-full of gasoline.

To save money, two-minute flares.
Lights must stay low. Never aim

a beam straight at a plane:
the pilot would be blinded.

Cruising at night under the stars,
a pilot can hold the ship steady,

watch the moon move.
A full moon hones landscape:

the knife-edge of the runway,
the grasses lining the verge,

the leaves shifting silver-sharp
on the oaks beyond.

With a wide moon in a clear sky,
the owner of a field may economize

for a few nights
and not deploy flares at all.

The Skywayman

Champ

Do a Tail Slide or a Dip of Death
in broad daylight, it's about
as dangerous
as scrambled eggs for breakfast.

Stunts after dinner are another matter.

The director of *The Skywayman* demands a mock-up.
Lock longs to look large, lives to do
hellacious things.

The director sighs. He'll shoot the night stunt—
the flaming crash—last.
Whatever the risks, they'll finish.
He paints the Jenny white,
orders bigger, brighter lights.

9:15. Lock drives to the field
with Viola. He's late.

The director positions five sun arcs
to pierce the night sky,
vertical columns.

9:40. Lock and Skeets take off,
ascend, level out,
cruise at three thousand,
loop in the dark space
between the lights.

10:05. Lock sets magnesium flares
on the wings, a cross of fire.
They dive to two thousand,

turn, spin counterclockwise,
spiral closer, closer.

Five hundred, time to pull up.

They spin lower,
lower—cut,
cut the lights!

CUT THE LIGHTS!

Hell's a-popping,
techs shift the sun arcs
straight at the cockpit.

Lock and Skeets slam into an oil sump,
ignite it.
Dry grass flames up
hell-west and winding.

Eternity Street

Champ

Lock the high god, the sky god,
and his sideman, Skeets, travel
in flag-fitted coffins
behind a marching band
and a parachute platoon.

Eighteen pilots sweep
across the sky in V-formations,
showering rose petals,

and twenty cowboy actors on horseback
escort two hundred cars
driving slowly,
single file.

Viola rides alone in a studio limo.

(Ruby's in charge at the church.)

Cameramen crowd Central Avenue.
Gaffers, grips, continuity girls,
seamstresses, stunt men, stunt women,
usherettes from the movie palaces,
and carnies from the carousel pier
fill porches and front yards.

Lock wing walked over red and yellow streetcars
scooting passengers on metal tracks.

He looped over men hammering
half-built houses in the hills,
soared above palms on new boulevards,
dove down on oil derricks nodding on dry flats.

He waggled his wings over swells of the metallic Pacific,
and he flew past tables at the Iowa picnic,
dropped a pitch onto a double-header,
buzzed a jazz concert, playing solos
with his wiry body hanging
from his huge hands.

Three miles long and a mile high,
my skywayman's funeral rolls
toward Eternity Street,
Calle de Eternidad,
and the air cortège hooks
double-banner headlines.

Faster than hell can scorch a feather,
Fox moves up the release.

III

Wedding in the Air

Pang

Here's our pitch: Come on up, fly
with the angels! We take you
high or low, fast or slow.

Pilots fake it in films, we carry
thousands who soar sitting next
to our sheepskin shoulders

and watch us loop,
spin, whipstall, wing-over,
wing walk. Every weekend

we add a wedding in the air,
a cross between a stunt
and a prayer:

I load a bride in tulle,
a groom in tails,
a minister to splice them:

breezy rites at cloud altitudes beat
marching down the nave in town
to an organ wheezing Mendelssohn.

The groom holds his hat, the couple
pays for a long, long ride, people
ogle vows sworn on the sky side

of the dollar. After the show, the boys and I
head for the juke joint on Airport Road,
Anywhere, U.S.A. (They name the road

before they build the airport.)
We do some ground flying.
You bet I stay single.

Q & A with Miss Law

Ruth

Q. Miss Law, how did you feel when you set your first record?
A. Triumphant. Icicles descended from my hair.

Q. And after that?
A. I launched the Ruth Law Flying Circus.

Q. What's on the bill?
A. Plane change in mid-air. I grope for a rope ladder. Miss.
   Miss again.

Q. Why?
A. Scares trigger bigger thrills: the spectators scream,
   groan as the ladder sways, gasp as I grab it at last.

Q. Have you ever failed?
A. I mount rung after rung, arm over arm, swim up the sky.

Q. Do you cheat?
A. I promise to defy gravity.

White Gate Blues

Bessie Coleman, pilot

I got the Waxahachie, Texas blues,
the colored schoolhouse, washtub, weevil blues,
a laundress has just one young life to lose,

I pound down on the clothes and soap them clean,
rinse out those piles of clothes and wring them clean,
I'm tenth of thirteen children, yes, thirteen.

Study some math at Colored Normal U.,
I love the math at Colored Normal U.
Run out of cash. Quit school. Try something new?

I long for flying lessons to lift me in the air,
I pray for flying lessons high up in the air,
God built no segregated space up there.

Give me lessons, I'll wing across the sky,
flying lessons, I'll cruise right through the sky,
God made no white supremacy on high.

Oh, Bessie, pilots tell me, no can do,
lessons, white men tell me, no can do.
Defender asks, Bess Coleman, *parlez-vous?*

And now New York and Coronado Bay,
Chicago, Kansas City, and LA—
and even Waxahachie—want to pay

to see me climb, and spin, and launch a loop,
to watch me roar above their heads and loop,
to see me grin a grin and whoop a whoop,

though I won't fly a Texas show unless,
no, I won't fly a hometown show unless
White Only signs come down before Queen Bess.

A pilot's got just one short life to lose,
A woman's got just one short life to choose.

Flying Lesson: Stunt Pilots' Price List

Blow up plane in midair, pilot parachutes out, $1,500
Crash plane, fly into trees, houses, etc., $1,200
Loop with person standing on wings, $450
Collide head-on with automobile, $250
Fight on wing, one person knocked off, $225
Fly upside down, $150
Change driver from motorcycle to plane, $150
Change from plane to plane, $100

Tomboy Stories

Ruth

Spare me Petticoat Pilot, Lipstick Flyer,
Winged Suffragette, Flying Flapper, Ladybird,

Bird Girl, Girl Hawk, and Tomboy of the Air.
Betty Scott ripped up that poster,

tomboy meant a loud, loose woman to her,
not someone up to her elbows in engine oil.

A reporter called me tomboy, too: slim, yellow hair,
blue eyes, looks like some high school girl.

I was twenty-five, a skilled pilot. I blame Pickens
for pitching Tiny Broadwick as Doll Girl,

Kate Stinson as Flying Schoolgirl. Champ
advised curls with ribbons, subtracted years.

Back when that reporter thought I should be
scribbling homework, I stunted

for a crowd at Narragansett, flew
in a drizzle. They handed me five hundred

damp, crumpled singles, way too little
for the risk. And the biggest risk? Men sabotage

women's planes—slash wires, drop a tool to jam
the controls. (When Bessie Coleman cracked up,

they found a wrench.) Watch out for the guy
who drains your gas tank, spikes your fuel,

claims it's a joke, claims women pilots have
no sense of humor. We can mark the ballot,

steer the ship. Air races are Men Only.
Time we tomboys got organized.

Attack on Lower Manhattan

Pang

Our fliers graze streetcars in St. Louis, snatch
      ornaments off steeples in Omaha, buzz
           water towers in Philadelphia.
                        I tell the boys,

let's stage an air war over Manhattan to help
      Colonel Billy Mitchell. He claims planes beat
           battleships, he tells the military planes can
                        threaten the U.S.A.

Our Red Standards loop the Battery, break away, roll,
      slip below the crown of the Woolworth Building, zoom
           above sidewalks on Broadway,
                      roar past

the windows of skyscrapers. New Yorkers take it all
      in stride. We gambol above Times Square,
           hover over trees in Central Park,
                      a million look up.

Traffic stops. We are famous! Tabloid photographers
      adore air war, *Evening Graphic* prints it all.
           We are getting rich! Mitchell is still
                  in trouble.

Flying Lesson: Bureau of Air Commerce

FOR IMMEDIATE RELEASE

1. No air shows over congested areas.
2. No pilot without a license.
3. No airplane without a registration.
4. No registration without an inspection.
5. No wing walker without a parachute.
6. No parachute jumper without a second parachute.

Flying Cars

Betty

A flock of names whispers lift: Autoplane, Tampier, Longobardi.
Flying cars hover above slow trucks—

they wing past stucco houses, commercial strips,
miniature golf, and the pale sand of the public beach. Like poets,

the designers test pitch and yaw. They practice gliding,
learn to wrap the extraordinary vertical in the everyday horizontal.

A flying car ascends, amazes, but like a figure of speech,
a roadable airplane must descend, drive on.

So, sell me a car to spring from the street, soar over traffic,
fold its wings, wheel home. Order me a fast, float-footed plane,

a flying boat to lift from a lake, cruise, splash down.
Build me a skyburb, each house with a hangar.

Light great circles, skyfaring routes. Swivel runways
into the wind, speed arrivals and departures.

Streamline everything in the U.S.A. Imagine 2020:
fifty thousand flights could take off every day.

Red Airplane

Pang

Slim Lindbergh strides knee-deep in ticker tape,
he invented mile-high celebrity—couples
at the Savoy swing the Lindy hop, leap

aerials every night, New York to Paris. Not for me.
Slim and I barnstormed Colorado long ago,
I can do distances too, long distances, fly

around the world in stages, four thousand,
five thousand miles nonstop.
Have to buy a Bellanca. Color it red,

like the Standards we flew in the circus.
Modify the tanks, carry extra fuel. Cut weight,
cut drag. Design a gadget to jettison

landing gear after take-off. Raise $110,000.
Pitch Hughie Herndon, his mother
could stake us. Name the ship *Alice*? And pitch

Veedol Motor Oil. Call the ship *Miss Veedol*?
Mark up the charts.
Leave my shoes behind.

Trans–Pacific

Champ

Hell's first whispers, I'd have won a bundle
if I'd put money on Pang this year.

That's old news.

He never signed with me, though I tried.
Pang's sky circus played every state

before regulators shut them down.

So, Pang ups his game, takes off with Hughie
from Japan, unlocks the landing gear, wing walks

over the Aleutians in gale-force winds, drops

the last two struts,
crosses the Pacific in forty-one hours,

comes in hellbooting it for a belly landing

in Wenatchee, Washington,
where—wait for it—his mother is watching.

His mother stitched the linen fuselage cover

for his first Jenny.
(What is it about some men's mothers?)

I'd be collecting a helluva bundle if

I'd laid money on Pang this year,
but I left aviation without a dime.

Catch me at the auto speedway,

killer track, thrill
a minute, slick straightaways

slide into banked curves.

Men will bet on anything
and I get a cut of everything.

Exuberance

Exuberance sips bootleg gin from a garter flask
with a ruby monogram, "E."

She wears a red dress one size too small,
eyes wide, she flirts with everyone,

dares Lincoln Beachey to run his tank dry,
ride a dead stick all the way down.

She watches Ormer Locklear climb
out of the cockpit two hundred feet up,

tap dance on his upper wing
as the houses of honest families

with their square-fenced yards
slide below his shuffle. An oval pond

winks in the sun, like a zero.
Exuberance challenges Clyde Pangborn

to master the Falling Leaf, ignore
the Graveyard Spiral, the Doom Loop.

She's a show-off like Ruth Law, she
speaks French like Bessie Coleman.

These aviators predict every American will fly,
Exuberance believes *Everybody Ought*

to Be Rich, John J. Raskob explains why
in the *Ladies' Home Journal.* She gets stock tips

from her manicurist, call loans from her broker,
buys Radio, Seaboard Utilities, Sears,

orders shares in investment trusts—why not?—
chain stores keep multiplying. Cars, trucks,

planes, houses, this nation is all about growth,
growth and leverage, look at the skyscrapers shooting up,

men rivet steel, floor after floor, high-speed elevators
spring through the cores, planes soar over them all.

Sherman Fairchild is making a million
selling aerial photographs of real estate.

Exuberance travels constantly, owns land
in Miami, Miami Beach, Coral Gables, Palm Beach,

she trades binders on lots five times over,
befriends a Mr. Charles Ponzi from Boston

who is raking in a bundle near Jacksonville.
Prices for sand and palms are sure to rise,

Glenn Curtiss is buying two hundred thousand acres
in Dade County, he's building towns.

But how do we know when irrational exuberance
has unduly escalated asset values?

Wall Street is wing walking, call it barnstormer
capitalism, soon the bankers and the brokers

will steal the aviators' lexicon, claim
their own tailspins, nosedives, crack-ups,

graph their own doomsday cycles, wonder
how everything blue-sky stayed up so long.

Exuberance buys more stock on margin,
volume runs high, the ticker tape

can't keep up. Higher. Higher.

Characters

Addie
Addie M. Tate (1869–1955). The postmistress in Kitty Hawk, North Carolina, she loaned Wilbur Wright her sewing machine so he could shorten the cloth wings of the Wright brothers' first glider.

Champ
William Hickman Pickens (1877–1934). Born in Mountain Home, Alabama, he raced bicycles, sold jukeboxes known as "iron entertainers," and promoted automobile racing before he became the promoter and publicity agent for many aerial performers including Glenn Curtiss, Linc Beachey, Betty Scott, Tiny Broadwick, and Ormer Locklear.

Linc
Lincoln J. Beachey (1887–1915). San Francisco's "Boy Aeronaut" steered a dirigible and worked as a mechanic for Glenn Curtiss before he became an exhibition pilot famous for his daring, "The Man Who Owns the Sky."

Betty
Blanche Stuart Scott (1884–1970). Born in Rochester, New York, she drove a car across the country (the second woman to do this) before becoming the first woman in the U.S. to fly. After a six-year career as a stunt pilot, she became an actress and writer working in film, radio, and television.

Harriet
Harriet Quimby (1875–1912). Born on a farm in Michigan, this New York journalist and drama critic was the first woman in the United States to earn a pilot's license and the first woman pilot to fly the English Channel.

Tiny
Georgia Ann Thompson Broadwick (1893–1978). Born in North Carolina, she left work in a cotton mill to join a traveling circus and jump from a hot air balloon with a parachute. She was the first woman to make a parachute jump from an airplane.

Lock

Ormer Leslie Locklear (1891–1920). A Greenville, Texas carpenter and mechanic, he became an army flight instructor, invented the wing walk, became an exhibition pilot, and then, with his partners, performed in Hollywood silent films.

Pang

Clyde Edward Pangborn (ca. 1893–1958). Born in Bridgeport, Washington, he grew up in lumber camps in Idaho. After jobs in logging and mining, and extension classes in engineering at the University of Idaho, he became an army flight instructor, barnstormer, and co-owner of the Gates Flying Circus. Nicknamed "Upside Down Pang," he specialized in flying upside down. He was the second pilot to wing walk after Locklear, and the first pilot (with copilot Hugh Herndon) to fly nonstop across the Pacific. Later he organized U.S. and Canadian support for the RAF in Britain in World War II, worked as a test pilot, and ferried aircraft all over the world.

Ruth

Ruth Bancroft Law (1887–1970). Born in Lynn, Massachusetts, she had the resources to purchase her own airplanes and hire instructors. She set distance and height records, traveled widely to perform, and established the Ruth Law Flying Circus.

Bessie

Elizabeth Coleman (1892–1926). One of thirteen children of an African American mother and an African American and Native American father, she grew up in Waxahachie, Texas, where she picked cotton and worked as a laundress. She migrated to Chicago where she worked as a manicurist and ran a chili parlor. With support from Robert S. Abbott, publisher of the *Chicago Defender,* she learned French and earned her license in France at a school run by the Caudron brothers, becoming the first African American pilot.

Notes

Joseph J. Corn's *The Winged Gospel: America's Romance with Aviation, 1900–1950* (New York: Oxford University Press, 1983) launched my interest in this era. Many other vivid histories include Tom D. Crouch, *Wings: A History of Aviation from Kites to the Space Age* (New York: W.W. Norton, 2003); Paul O'Neil, *Barnstormers & Speed Kings* (Alexandria VA: Time Life Books, 1981); Robert Wohl, *A Passion for Wings: Aviation and the Western Imagination, 1908–1918* (New Haven: Yale University Press, 1994); David T. Courtwright, *Sky as Frontier: Adventure, Aviation, and Empire* (College Station, TX: Texas A & M University Press, 2005); Bernard Marck, *Women Aviators: From Amelia Earhart to Sally Ride, Making History in Air and Space* (Paris: Flammarion, 2009); and Eileen F. Lebow, *Before Amelia: Women Pilots in the Early Days of Aviation* (Dulles, VA: Brassey's Inc., 2002).

Poet Diane Ackerman's memoir, *On Extended Wings: An Adventure in Flight* (New York: Atheneum, 1985), captures the magical experience of learning to fly. The Smithsonian National Air and Space Museum in Washington, DC, the Museum of Flight in Seattle, the Cradle of Aviation Museum in Garden City, NY, and several others display some of the earliest flying machines. Biplanes made in the United States included the Wright Flyer, the Curtiss "pusher," and the Curtiss Jenny (JN-4D). Blériot monoplanes were imported from France. I use modern spelling (airplane not aeroplane) and terminology (airport not aerodrome).

A few books and articles about the characters are listed in the notes below.

Kitty Hawk, 1900 "We came...got them," letter from Orville Wright quoted in Peter L. Jakab, *Visions of a Flying Machine: The Wright Brothers and the Process of Invention* (Washington and London: Smithsonian Institution Press, 1990), 92.

Dominguez Field Air Meet Los Angeles, 1910. Ballyhoo or bally: carnival barker's pitch before an event. Twenty-four sheet: a billboard-size advertisement.

Bird Woman "The Car, The Girl, and The Wide, Wide World," painted on the side of her Overland car, Julie Cummins, *Tomboy of the Air: Daredevil Pilot Blanche Stuart Scott* (New York: Harper Collins, 2001), 10.

Suiting Up "I have the air intoxication . . . means," Matilde Moisant, Harriet's friend and fellow student, quoted by Elizabeth Hiatt Gregory, "Women's Record in Aviation," *Good Housekeeping* (September 1912): 145–149, reprinted in *Harriet Quimby, America's First Lady of the Air*, ed. Ed Y. Hall (Spartanburg, SC: Honoribus Press, 1990).

Drag ". . . that cub, Beachey," cited in "Los Angeles Meet Closes a Thorough Success," *Aero: America's Aviation Weekly* II (February 10, 1912): 376.

Daredevil Undertaker's Chair: the exposed seat in a Curtiss pusher.

Flying Lesson: Winds Terms for wind speed adapted from the 1806 Beaufort scale by Sir Francis Beaufort.

Accident "Certainly not . . ." and "They were going . . ." adapted from "Miss Quimby Dies in Airship Fall," *New York Times,* July 2, 1912, https://timesmachine.nytimes.com/timesmachine/1912/07/02/100541622.pdf (accessed May 12, 2018); and Lawrence Goldstone, *Birdmen: The Wright Brothers, Glenn Curtiss, and the Battle to Control the Skies* (New York: Ballantine, 2014), 329–330.

Flying Lesson: The Aviation Recipe adapted from Hugo R. Ensslin, *Recipes for Mixed Drinks* (New York: 1916).

Packing the Parachute "You're awful small . . ." Elizabeth Whitley Roberson, *Tiny Broadwick: The First Lady of Parachuting* (Gretna, Louisiana: Pelican, 2001), 58; "Honey, that's the way . . ." 14; "this was all . . ." 16.

Panama-Pacific "The medal is not here yet; could you go up one more time?" Frank Marrero, *Lincoln Beachey: The Man Who Owned the Sky* (San Francisco: Scottwall Associates, 1997), 174. Jump rope rhyme, adapted from 187.

The Wing Walker "Flying circus" first coined by Pickens with Curtiss and Beachey in 1911. "Flying circus three miles long and a mile high," Art Ronnie, *Locklear: The Man Who Walked on Wings* (South Brunswick, NJ and New York: A. S. Barnes, 1973), 57.

Stunt Money ". . . rain, shine, or cyclone," and "Bandages are box office," Ronnie, *Locklear*, 157 and 79.

The Barnstormer "Rides . . ." and "Let's Get Acquainted," air circus handbills in the Lindbergh Papers, Manuscripts and Archives, Sterling Library, Yale University.

Flying Lesson: Strange Field Landings On contact flying, see Dean C. Smith, *By the Seat of My Pants: A Pilot's Progress from 1917 to 1930* (Boston: Atlantic Monthly Press/Little Brown, 1961), 106–107.

Feel "The scare is part of the thrill," newspaper interview quoted in Heather Lang, *Fearless Flyer: Ruth Law and Her Flying Machine* (Honesdale, PA: Calkins Creek, 2016), 15. Also see "Ruth Law—Queen of the Air: Challenging Stereotypes and Inspiring a Nation" by Billie Holladay Skelley, www.ninety-nines.org/ruth-law.htm (accessed April 5, 2017).

Shroud Lines "I drift down . . . grass," details adapted from Bill Rhode, *The Flying Devils: A True Story of Aerial Barnstorming* (New York: Vantage Press, 1983), 139–140.

Heaven ". . . how much . . . there," adapted from Charles A. Lindbergh, *"WE"* (New York, G. P. Putnam's Sons, 1927), 60. Early in his career, Lindbergh was a barnstormer like Pangborn.

Safety Second "Safety second is my motto," Ronnie, *Locklear*, 9.

The Skywayman On the accident, see Ronnie, 243–278. The lights blinded the pilots; they may have melted the control wires as well.

Eternity Street Eternity Street is the road to the cemetery on the Ord map of Los Angeles.

Wedding in the Air Ground flying: talking about flying while on the ground. On the circus, see Bill Rhode, *Baling Wire, Chewing Gum, and Guts: The Story of the Gates Flying Circus* (Port Washington, NY: Kennikat Press, 1970), and Carl M. Cleveland, *"Upside-Down" Pangborn, King of the Barnstormers* (Glendale, CA: Aviation Book Company, 1978).

White Gate Blues Robert Abbott, publisher of the *Chicago Defender*, helped support Coleman when she went to France for flying lessons. See Doris L. Rich, *Queen Bess: Daredevil Aviator* (Washington, DC: Smithsonian Institution Press, 1993), 94–95 on the gates. In Waxahachie, Coleman insisted on one shared gate but blacks and whites were still seated separately.

Flying Lesson: Stunt Pilots' Price List Adapted from Bon McDougall, 13 Black Cats, quoted in O'Neil, *Barnstormers*, 53.

Tomboy Stories Blanche Scott, "Tomboy of the Air," poster from 1912, Lebow, *Before Amelia*, 43. "... yellow hair ..." and "There's too little money for the risk," Lebow, *Before Amelia*, 201. For Law's ten-year career, Lebow, *Before Amelia*, 201–224. Women pilots eventually organized as the Ninety-Nines in 1929. Louise Thaden won the Bendix Trophy air race in 1936; Jacqueline Cochran won it in 1938.

Attack on Lower Manhattan Colonel Billy Mitchell campaigned for more military funding to support fliers in future air wars and was court-martialed for speaking out too strongly.

Flying Cars Glenn Curtiss developed an experimental flying car as well as the Curtiss N-9, a successful flying boat. Later designs for flying cars were more effective and prototypes performed well, though none achieved mass production. For flying boats, cars, skyburbs, and futuristic airports, see Corn, *Winged Gospel*, 58 and 91–111, and www.roadabletimes.com.

Red Airplane According to some dance historians, the Lindy hop was named for Charles A. "Slim" Lindbergh (also called "Lindy"), after he flew the Atlantic. Cutting weight saved fuel. Lindbergh cut out the parts of charts he did not need; Pangborn ditched the landing gear and left his shoes behind.

Exuberance "But how do we know..." is from a speech by Alan Greenspan, quoted by Robert J. Shiller, *Irrational Exuberance* (Princeton: Princeton University Press, 2000).

Acknowledgments

Thank you to the Djerassi Resident Artists Program, the Martha's Vineyard Writers Residency, the Fine Arts Work Center, and the Virginia Center for the Creative Arts for time and space to pursue this project, and to Yale University for leave. Many writers (and a few pilots) read poems and offered encouragement, including Janet Bednarek, Eavan Boland, Harriet Scott Chessman, Holly Coddington, Martha Collins, Robert Goyer, Catherine Graham, Alice Kaplan, Delaney Lundberg, Barbara Marks, Laura Marris, Gardner McFall, Leslie Monsour, Ginger Murchison, Marilyn Nelson, Katha Pollitt, Will Schutt, Tom Sleigh, Pat Valdata, Marc Vincenz, and Reed Wilson. Thank you as well to Rabia Ali, Patty Baldwin, Ellen Bogart, Kim Bridgford, Joshua Chuang, Barbara Henry, Alice Pentz, Diane Thiel, and Jock Reynolds who provided opportunities for me to read from work in progress. I'm grateful to Eloise Klein Healy and Colleen Rooney, who introduced me to Red Hen Press, and to Kate Gale and everyone who worked on *Exuberance*.

Special thanks to editors Don Share, Lindsay Garbutt, Anna Lena Phillips Bell, J.D. McClatchy, Susan Bianconi, Jackson Lears, Stephanie Volmer, Bill Ray, Joyce Wilson, Scott Wiggerman, Cindy Huyser, Sarah Pemberton Strong, Willard Spiegelman, Jennifer Cranfill, Lori Desrosiers, and Hannah Sassoon who published poems, often in earlier versions:

American Scientist, "Flying Cars"; *The Courtship of Winds*, "Daredevil," "Postures," "Panama-Pacific" (as "Loop After Loop"); *Bearing the Mask: Southwestern Persona Poems* (Dos Gatos Press, 2016), "The Wing Walker"; *Ecotone*, "Kitty Hawk, 1900" (as "1900"), "Bird Woman" (as "Bird Woman, 1910"); *Naugatuck River Review*, "Tomboy Stories"; *New Haven Review*, "Packing the Parachute," "Flying Lesson: Emergency Jump" (as "Flying Lesson: Air Mail, 1920"); *Poetry*, "Exuberance," "Flying Lesson: Clouds" (as "Flying Lesson," also a letterpress broadside from the Center for Book Arts, and a Poetry Out Loud selection); *The Poetry Porch*, "Flying Lesson: Winds," "White Gate Blues"; *Raritan: A Quarterly Review*, "Flying Lesson: The Aviation," "Flying Lesson: Stunt Pilots' Price List," "Flying Lesson: Bureau of Air Commerce," "Heaven," "Playing the Pastures," "Shroud Lines," "Attack on Lower Manhattan" (as "Selections from Eternity Street"); *Southwest Review*, "Eternity Street"; *These Fifty States*, "Landscapes"; and *The Yale Review*, "Bird Man" (as "Bird Man, 1911").

Biographical Note

Dolores Hayden, award-winning poet and historian of American land-scapes, investigated the lives of daredevil pilots—women and men from the earliest years of aviation—for *Exuberance*, her third poetry collection. Hayden's poems have appeared in *Poetry*, *The Common*, *Ecotone*, *Raritan*, *Shenandoah*, *The Yale Review*, *Southwest Review*, *Best American Poetry*, and *Poetry Daily*. Author of *American Yard* (2004) and *Nymph, Dun, and Spinner* (2010), she's received awards from the Poetry Society of America and the New England Poetry Club, and residencies in poetry from Djerassi, the Virginia Center for the Creative Arts, and Noepe. Professor of Architecture and American Studies Emerita at Yale University, Hayden has also been a Guggenheim Fellow and won an American Library Association Notable Book Award for nonfiction. Her website is www.DoloresHayden.com.

CPSIA information can be obtained
at www.ICGtesting.com
Printed in the USA
JSHW030811140922
30398JS00005B/18

9 781597 096041